Chilton County Peaches

Lynn Edge

Seacoast Publishing, Inc.
Birmingham, Alabama

Chilton Co. Peaches

Published by
Seacoast Publishing, Inc.
P.O. Box 26492
Birmingham, Alabama 35260
(205) 979-2909

Copyright © 1994 by Lynn Edge

All rights reserved.
We are delighted for reviewers and writers of magazine and newspaper articles to quote passages from this book as needed for their work. Otherwise, no part of this book may be reproduced or transmitted in any form or by any means, electronic or mechanical, including photocopying, recording or by any information storage and retrieval system, without written permission of the publisher.

ISBN 1-878561-33-2

Designer Lori Leath-Smith

To obtain copies of this book by mail, please send $6.50 per copy to Seacoast Publishing's address, listed above. This price includes shipping and handling.

For my grandmother and grandfather—Fairy and Clarence Hall—who flavored my childhood memories with peaches and filled my life with love. And for Becky, who shared the summers, the peaches and the love.

"He who eats the peach...is first of all agreeably struck with the perfume which it exhales; he puts a piece in his mouth, and enjoys a sensation of tart freshness which invites him to continue; but it is not until the instant of swallowing, when the mouthful presses under his nasal channel, that the full aroma is revealed to him; and this completes the sensation which a peach can cause."

—*Anthelme Brillat-Savarin, gourmet, giving tasting instructions for eating a peach*

INTRODUCTION

There probably aren't too many people who grew up in the South who don't have what I like to think of as "peach memories"—some stories of childhood written in terms of peaches. For some people the stories might be tales of picking and eating fresh ripe peaches. For others, they might be adventures of climbing the trees. For still others, they might be mischievous stories of the joy of pinging a younger brother or sister with a hard, green peach just when it was least expected. Whatever the story, most of us grew up with peaches just because we were fortunate enough to live in an area of the country where they are grown in orchards and in back yards.

My peach memories are of sitting on the porch of my grandmother's home on summer afternoons. There would be a huge wash tub filled with water and peaches at the beginning of the afternoon. At the end, there would be a washtub filled with water "itchy" with peach fuzz and pans and pans of peeled and sliced peaches for eating and for canning. In between, there was time to talk to my grandmother as she reached again and again into the tub for another peach to peel. We made up stories, we talked about when she was growing up and we ate more than a few slices of peach as the day went on.

Now I'm married to a man who loves Alabama peaches. He doesn't grow them, but he eagerly anticipates the harvest each year and spends much of the winter worrying whether the temperature and other conditions have been right for producing a big crop that season.

I suspect he's not the exception in Alabama, but more like the rule. We all wait for reports from Chilton County on how the peaches are doing. And each year, when the peaches finally are "in," we are reminded why they say in Chilton County that these are the best peaches in the world.

Chilton Co. Peaches

Table of Contents

1
The history of peaches
in Chilton County and elsewhere 1

2
Varieties of peaches 7

3
Peach how-tos 13

4
How much/how many/how good for you? 17

5
The perfect peach—
what do you do with it? 19

6
Celebrating Chilton County's peaches 27

7
Cooking with peaches 29

Chilton Co. Peaches

1
The Peach - A Brief History

Just about everyone who has grown up in the South has a wonderful summer memory—biting into a peach and having that sweet, delicious flavor fill the mouth, sticky juice running down the chin. That memory, that taste is a treasure that seems so Southern, yet those who research the history of plants tell us it is a gift from other lands.

The Romans called the peach *Prunus persica*, meaning they thought it came from Iran, according to Edward Hyams' Plants in the Service of Man. Hyams says there's no evidence that the peach existed in ancient Greek gardens and that it only came to Italy in the First Century A.D.

But, he points out, the Romans must have been wrong about just how it got to them because the peach "is not and never has been native" to Iran. The only plant close to a "wild" peach—the forefather of the ones we enjoy today—has been found in China, where it was grown at least 4,000 years ago. There are Chinese writings containing references to the peach as early as the Fifth Century B.C.

Chilton Co. Peaches

The peach must have been "born" in China, Hyams feels. The fruit of the peach tree was a symbol of immortality to the ancient Chinese. They exchanged the fruit - both real and porcelain peaches - as a token of affection and placed bowls of peaches in the tombs of loved ones to prevent the bodies from decaying.

In 1977, archaeologists in China discovered the perfectly preserved body of the wife of the Marquis of Tai. She had been buried in the Second Century B.C. In the tomb, they found a bowl containing peaches. Since the early Chinese dynasties, the people had believed that peaches, eaten at the proper time before death, would preserve the body "until the end of the world."

Explorer Marco Polo once reported that upon visiting China (in fact, in the area of the country where the woman's body was later discovered), he had seen very large peaches "weighing fully two small pounds apiece."

Under one of the Chinese dynasties in the Second Century B.C., the country exchanged useful plants and animals with Iran. That's probably how it got to that country and from there, it was introduced to Greece and, later, to Rome.

Once the fruit, which the Chinese had domesticated from the wild plants, got to Italy, it wasn't any time before it spread all over Western Europe. Because the peach is a very adaptable plant, Hyams says, it must have flourished in its new European home.

Like many new things that come along, the peach at first may have been a delicacy reserved for the meals of the well-to-do and the aristocratic.

In Europe's "Low Countries," according to Reay Tannahill's Food in History, much of the food was heavy and fatty in the 1600's, but the Dutch had begun to cultivate as many "exotic" fruits as they could get to grow. In fact, by 1636, still life artist Jan Davidsz de Heem moved to Antwerp because "there one could have rare fruits of all kinds, large plums, peaches, cherries, oranges, lemons, grapes and others," to use as "models" for his artwork. The fruits there, he said, were "in finer condition and state of ripeness to draw from life."

In England at the same time, fruits were only on "the tables of the great, and of a small number even among them," according to Tannahill. One English gardener of that era boasted that there were four varieties of peaches - white, red, yellow and d'avant - and "I have them all in my garden, with many other sorts."

The peach probably came to England from France. William Lawson, the English author of A New Orchard and Garden, written in 1618, warned English gardeners: "Meddle not with Apricoekes nor Peaches...which will not like our cold parts unless they be helped with some reflex of Sunne."

Brought to North America by the Spaniards (It had been independently introduced from the Near East to Spain.) in the 16th century, the peach was planted all over the New World by Native Americans and colonists as well. Once again, the adaptable peach spread quickly. Because the plant got along so well without human help, the trees soon were everywhere.

The Native Americans seemed to immediately fall in love with the new plant. While many tribes, especially the Creeks and the Seminoles, grew the trees, the Natchez Indians even named one of the months after the fruit.

The quick spread of the peach plant may have confused some early North American residents. Explorers who found peaches in places were only Native Americans lived thought the plant had to be native to the New World since no Europeans could have brought it to that area.

In the 1600's, William Penn found peach trees in native gardens in New England, writing in a letter, "There are...very good peaches (in Pennsylvania) and in great quantities; not an Indian plantation without them...not inferior to any peach you have in England, except the Newington."

One of the most popular and well-known of peach varieties - the Elberta - is something of a southern creation. In 1850, the story goes, the Rumphy family of Georgia got some peach buddings from China. They planted them and the trees produced an abundance of fruit.

Mrs. Rumphy is supposed to have dropped several of

Chilton Co. Peaches

the pits into a basket and forgotten about them until her grandson, interested in starting an orchard of his own, found them. He planted the pits and the trees once again flourished. Cross-pollination took place between the older Chinese trees and Rumphy's newer ones and a new peach was born.

Rumphy named the peach for his wife - Elberta.

The Peach and Chilton County

Chilton County, located in the central part of Alabama, contains about 700 square miles and is the 31st largest county in the state. The county lies in a part of the state that has mild winter and warm summers. It's a place where the sun shines an average of 59% of the daylight hours.

It's a place where peaches grow.

According to Chilton County and Her People, written in 1940 by T.E. Wyatt, then editor of the Union-Banner, raising peaches in Chilton County "just come natural. Our soil and our climate are simply ideal. Old Mother Nature saw fit to bless us that way."

Wyatt says there is evidence that peaches were in what would become Chilton County when the first settlers got there. History, he says, tells the story of how Hernando DeSoto and his group of explorers came down the Coosa River in the 1500's. At that time, Wyatt says, there was an Indian town, Pokana Talahassi, in the area. In the Muskogee Indian language, he says, "Pokana Talahassi" meant "Old Peach Tree Town."

According to Wyatt, one of the first people to raise peaches commercially in Chilton County was P.C. Smith, who came to the area from Georgia. Wyatt says a Mr. and Mrs. Martin Peterson, natives of Denmark, were the first to set out peach trees in Thorsby. They had come to the small town from Iowa in 1898.

> The ancient Chinese considered it unwise to plant a peach tree near the bedroom window of a young girl because the intoxicating aroma might cause her to have impure thoughts.

Because the peach business is big business in Chilton County, it is natural that the people there celebrate the fruit and the harvest. This celebration, the Chilton County Peach Festival, began in 1947, Wyatt said in his revised edition of Chilton County and Her People, published in 1950. That year, the peach crop in Chilton County brought the growers more than $2 million. There were then about 700 acres in peaches in the county covered with around 7,000 trees.

Agriculture remains an important part of the county economy. In fact, in 1990, it was ranked as one of the county's major industries. One of the primary fruits grown there, of course, is peaches. (Fifteen million pounds of peaches were produced in Alabama in 1989 and a great number of those were grown in Chilton County.)

Chilton Co. Peaches

2
Peach varieties

All peaches are alike in some ways. They have a slightly furry, downy skin on the outside and a single seed or "pit" on the inside.

And while all kinds of peaches share those qualities, there are differences from peach to peach. Some are yellow-fleshed while others have white flesh (These varieties usually have an almost berry-like aroma.). Some are freestone and others clingstone - catagorized on the basis of how tenaciously the flesh adheres to the pit.

Among the first modern peach varieties was the Elberta, the ancestor of today's yellow-fleshed freestone peach. It was developed in Georgia hundreds of years ago and has become one of the most popular peaches around the world today. In fact, at one time, the Elberta was so popular that it alone accounted for more than 90 percent of the total peach output in the United States. Though other varieties have increased in popularity and taken away some of the Elberta's clout, that peach still makes up about 50 percent of America's peach crop each year.

Since the Elberta was developed, a number of other

Chilton Co. Peaches

> Thomas Jefferson planted European peach stones on the grounds of his home at Monticello in 1802. The seeds had been sent to him by Philip Mazzei, his friend and former neighbor in Virginia. Mazzei returned to his native Pisa, Italy, and sent Jefferson the seeds from peaches grown there.

peach types have come along. Today, there are hundreds of supermarket varieties. Generally, the newer varieties are larger, firmer and have a more acid flavor than the older types.

In Chilton County, peach production has been an important industry for many years. While there are more than 75 varieties being grown throughout the county, there are some types that have been grown by most producers for a number of years. The names of these well-known varieties have come to mean "peach season" to those who enjoy the fruits of the Chilton County peach orchards.

The major varieties grown in the county include:
Camden. A small, early clingstone peach released in 1972. The Camden peach ripens around May 20th, making it the earliest ripening peach in the Southeast for many years now.
Junegold. A golden-yellow oval peach with a firm, juicy, aromatic flesh, the Junegold is a clingstone fruit. This variety was released in 1958. A large fruit for the season, it ripens around June 6th.
Dixired. A medium-sized, yellow, freestone peach, this variety was developed in America and has been around since 1945. The peach, which ripens around June 12, has a firm, very juicy flesh.
Coronet. A semi-clingstone peach with pale to dark yellow flesh, this variety is soft, juicy and fragrant. It, too, was developed in America and was released in 1953. A very popular variety in the Southeast, the Coronet ripens around June 18.

RedHaven. A medium to large, yellow, semi-clingstone peach, RedHaven has a very firm, juicy, aromatic flesh. Released in 1940, it is one of the most widely grown peach varities of all times. This variety ripens around June 22.

Harvester. A consistent producer, this variety yields an attractive, freestone fruit. Released in 1973, Harvester ripens around June 28.

Redglobe. This freestone peach shows up just in time to be part of one of the summer's biggest celebration. The Redglobe variety, released in 1954, ripens around July 4.

Loring. This large, yellow, freestone peach is a very popular and well-known variety. Released in 1946, the Loring ripens about July 6.

Cresthaven. A very large, yellow, freestone peach with firm flesh, Cresthaven was released in 1963. It ripens around July 12.

Redskin. A large, yellow, freestone peach with firm flesh, this variety was released in 1944. It ripens around July 15.

Elberta. This popular variety yields a large, yellow, freestone peach. The Elberta, released in 1889, ripens around July 18.

While these varieties still are around, they are giving way slowly to newer types.

Other Alabama varieties you may encounter include:
Sentinel. A medium, white, semi-freestone peach.

Ranger. A medium, yellow, freestone peach with firm flesh.

Blake. A large, yellow, freestone with firm flesh.

Rio-oso-gem. A very large, yellow, freestone peach.

Other peach varieties found in various parts of the United States

> Europeans like to slice a fully ripened peach into a glass of wine at the beginning of a meal. When the meal is finished, the peach slices are eaten as dessert.

Chilton Co. Peaches

include:
Babcock. A medium, white, semi-freestone peach.
Belle of Georgia. A large, white, freestone peach.
Biscoe. A large, yellow, freestone peach with firm flesh.
Burbank July Elberta. A large, yellow, freestone peach with firm flesh. This very old variety has become a favorite. It was developed by Luther Burbank.
Com-Pact RedHaven. A medium to large, yellow, freestone peach with very firm flesh.
Culinan. A large, yellow, freestone peach with firm flesh.
Cumberland. A medium, white, semi-clingstone peach.
Dawne. A medium, yellow, freestone peach.
Delp Early Hale. A very large, yellow, freestone peach.
DesertGold. A medium, yellow, semi-clingstone peach with soft flesh.
Early RedFe. A medium, white, semi-freestone peach.
Early RedHaven. A medium, yellow, semi-clingstone peach.
Garnet Beauty. A medium, yellow, semi-freestone peach.
GloHaven. A large, yellow, freestone peach.
Golden Jubilee. A medium, yellow, freestone peach.
Golden Monarch. A freestone peach with firm flesh.
Harbinger. A small to medium, yellow, clingstone peach with soft flesh.
Harken. A large, yellow, freestone peach.
Havis. A large, yellow, freestone peach.
J.H. Hale. A very large, yellow peach with firm flesh.
Jerseyland. A very large, yellow, semi-freestone peach.
Madison. A yellow, freestone peach with firm flesh.
Marsun. A large, yellow, freestone peach.
Maybelle. A medium, white, semi-clingstone peach.
Monroe. A large, yellow, freestone peach with firm flesh.
Norman. A large, yellow,

> The peach is Number 3 on the list of favorite foods, led only by apples and bananas.

freestone peach with firm flesh.
    Raritan Rose. A large, white, freestone peach.
    Reliance. A medium, yellow, freestone peach.
    Springcrest. A small to medium, yellow, semi-freestone peach with firm flesh.
    Stark Autumn Gold. A very large, yellow, freestone peach.
    Stark Earliglo. A large, yellow, freestone peach.
    SunHaven. A large, yellow, freestone peach.
    Sunhigh. A large, yellow, freestone peach with firm flesh.
    Sunshine. A large, yellow, freestone peach.
    Topaz. A very large, yellow, freestone peach.
    Triogem. A large, yellow, freestone peach.
    Velvet. A medium to large, yellow, freestone peach with firm flesh.
    Washington. A large, yellow, freestone peach with firm flesh.
    Waverly. A medium, white, semi-clingstone peach.
    Winblo. A large, yellow, freestone peach with firm flesh.
    Yakima Hale. A large, yellow, freestone peach.

Chilton Co. Peaches

3

Peach how-tos

Choosing peaches

In the supermarket, choosing the perfect peach is a lot easier that picking out the right watermelon. You don't have to thump them, lift them or take any of the other elaborate steps watermelon selectors have devised.

While many people may suggest various "feeling" methods of choosing peaches, almost all of them agree that smell is the best sense to use when selecting peaches. If it smells "peachy," it probably tastes peachy.

If you are one of those who has to have some "hands-on" experience when choosing fruit, all you need to do is look and gently feel. The peaches you choose should be fairly firm to slightly soft. There should be deep red areas on the skin surrounded by light yellow skin with a creamy tint to it. Give a gentle squeeze around the stem. It should give just a little and not be rock hard.

Very firm or hard peaches may be immature and may not ripen properly. Peaches with large brown spots on the skin or those that show signs of decay aren't the ones to choose, either.

If you are growing your own peaches, you can learn to

Chilton Co. Peaches

> "A French peach is juicy and when you first bring it in contact with your palate, sweet, but it leaves behind a cold, watery, almost sour taste. It is for this reason so often eaten with sugar. An American is exceedingly apt to laugh if he sees ripe fruit of any sort eaten with anything sweet. The peaches here leave behind a warm, rich and delicious taste, that I can only like in its effects to that which you call the bouquet of a glass of Romanee."
>
> —*The Travelling Bachelor by James Fenimore Cooper*

tell if they are ripe on the tree.

The Chilton County Extension Service recommends that you look at the undercolor, not at the red blush. On peach varieties with yellow flesh, the undercolor changes from green to light green to yellow. On those with white flesh, the color will change from green to light green to ivory.

For maximum flavor, pick the peaches when all of the green color is gone from around the stem end.

Picking peaches from trees

To pick a peach without bruising it or damaging the tree, cup the fruit in your hand and lift with a slight twist. The short stem will separate cleanly.

Ripening peaches

If the peaches you've picked haven't quite gotten ripe enough, there's an easy solution—let them continue ripening off the tree.

Ripen peaches by storing them at room temperature in a brown paper bag. (The ripening process stops once the peaches are put in the refrigerator.) Some experts recommend putting a nearly ripe apple or banana in the bag with the peaches.

Check the peaches daily to see how ripe they've gotten.

When they are as ripe as you want them, put them in the refrigerator to preserve that taste you've been waiting for. Eat them within one or two days because peaches ripened this way lose their flavor quickly.

Pitting whole peaches

To pit a whole peach, cut an almond shape around the stem end through to the stone. At the blossom end, poke a potato nail or skewer into the peach and push the stone out through the cut.

> REMEMBER: Unlike apples and pears, peaches do not improve in sugar content or ripen well off the tree. For the best flavor, let them ripen before they are picked.

Pitting and slicing peaches

To pit and slice a freestone peach, slice around the "seam" and twist the peach in half then lift or cut out the pit. The fruit of a clingstone peach has to be cut away from the pit in quarters or slices.

Peeling peaches

To remove peach peels the quick and easy way, drop the whole peach into boiling water for 10 to 30 seconds (The riper the fruit, the shorter the time.) just before you're ready to use them. Remove them with a slotted spoon and drop them in cold water. The skins should slip right off.

Chilton Co. Peaches

4

How much/How many/ How good for you?

One pound of peaches is about 3 medium-sized or 2 large peaches. This amount of fruit yields 1½ to 2 cups of sliced fruit and 1½ cups of puree.

When freezing peaches, one bushel (about 48 to 50 pounds of fresh peaches) will fill 18 to 24 quart containers. Two and one-half pounds of fruit will yield about 1 quart.

When canning, one bushel of peaches will yield about 10 to 24 quarts. (One quart is about 2 to 2½ pounds of peaches.)

> In the court of Louis XIV, the peach was called "teton de Venus," the breast of Venus

Peach Nutrition

A fresh Alabama peach offers two major types of nutrients—Vitamin A and Vitamin C.

Chilton Co. Peaches

> When peaches first were brought to North America, they were used for feeding hogs and for making brandy and many other strong spirits.

Two medium peaches provide one-half the minimum requirement for these two vitamins and each peach has only about 38 calories.

When eaten with the skin on, the peach also provides about 3 grams of fiber.

5

The perfect peach - what do you do with it?

Once you've located and acquired the perfect Chilton County peach, what do you do with it? You might think the answer is obvious - you eat it, of course - but what if you want to save that flavor for later, say months later?

You can do that, too, by freezing, canning or pickling the peach. Here's some advice from the experts on how to best enjoy the Chilton County peach now or later:

EAT THEM NOW

Enjoy fresh peaches just as they are, chilled or at room temperature, the Chilton County Extension Service says. You can wash and eat them peel and all, just like an apple. Or you can peel them first. If you do remove the peel, always peel just before serving to keep the color bright. If you have to peel them ahead of time, coat them with orange or lemon juice to keep them from turning brown.

FREEZE THEM

Chilton County peaches also are great for freezing, according to the Extension Service. When you plan to freeze

Chilton Co. Peaches

> One of the most famous peach desserts—Peach Melba—was created in the late 19th century to honor Dame Nellis Melba.

the peaches, select fruit that is fresh, mature and firm-ripe. Freeze immediately or refrigerate them until you can freeze them.

To ready them for freezing, wash peaches then remove peel and seed. Next, cut the peaches up in the proper size segments for however you plan to serve them when they are thawed.

Drop into anti-darkening solution (See examples below.), then freeze.

It's best to prepare two quarts, then pack into containers and cover with syrup or sugar mixture whichever you prefer (See "Types of Packs" below.). This gives you about the right amount of peaches to be working with at one time so you get them "put up" before they start to turn brown.

When you put the peaches into the freezing containers, allow about ½ to 1 inch of "head" space to allow food to expand as it freezes. If freezer bags are used, be sure to remove the air from the bag before sealing.

Finally, seal and label the containers.

Now place the containers in the coldest part of your freezer and freeze quickly.

Anti-darkening solutions

Peaches turn brown once they are exposed to the air. It's best to peel a small amount of peaches at a time and put peeled fruit immediately into one of the following solutions, according to the Extension Service:

1. To one gallon of water, add 2 tablespoons of salt and 2 Tablespoons of distilled (clear) vinegar. Do not let fruit stay in too long. It will absorb vinegar and salt. Rinse fruit before adding sugar or syrup.

2. To one gallon of water, add 1 teaspoon citric acid. Fruit does not absorb citric acid and does not need to be rinsed before adding sugar or syrup.

Chilton Co. Peaches

3. To one gallon of water, add 1 teaspoon ascorbic acid. Fruit does not need to be rinsed.

4. Use ascorbic acid mixture, following manufacturer's directions.

Types of packs
Peaches that are packed in sugar or syrup keep their flavor, color, texture and aroma longer and do not turn dark as quickly after defrosting, the Extension Service advises.

Peaches, however, can be packed and frozen without sugar.

1. Sugar pack. Add ascorbic acid mixture to sugar (1 cup sugar, plus ascorbic acid or ascorbic acid mixture, will cover 4 to 6 cups of fruit) according to container directions. Coat fruit well with sugar mixture, pack into containers and freeze.

2. Syrup pack. A 40% syrup (3 cups sugar in 4 cups water) is recommended for peaches. To use ascorbic acid mixture, add mixture according to container directions to cold syrup. To use ascorbic acid, dissolve sugar in lukewarm water and add ¼ teaspoon acid for each cup of syrup. Allow ½ to ⅔ cup syrup for each pint of fruit. Pack peaches in moisture-vapor proof containers, cover with syrup, remove air, seal, label and freeze.

3. Light syrup. (This has fewer calories than sugar or syrup packs.) Use 1¼ cups sugar with 4 cups water. Add ascorbic acid or ascorbic acid mixture. For extra sweetness, add some low-cal sweetener to cooled syrup. Sweeteners added to cooled, light syrup will not turn bitter in the freezer as they might in canning.

> Madame Recamier, a famous beauty of the early 19th century, is said at one time to have been so ill that we refused all foods. Her attendants, afraid she was dying, tempted her into eating by offering her a dish of peaches and cream. Madame Recamier recovered fully.

Chilton Co. Peaches

What other people call the peach:
French: peche
German: Pfirsich
Italian: pesca
Spanish: melocoton
Portuguese: pessego
Danish/Norwegian: fersken
Swedish: persika
Finnish: persikka
Russian: persik
Polish: brzoskwinia
Serbo-Croat: breskva
Romanian: piersica
Bulgarian: praskova
Greek: robakinon
Turkish: seftali
Hebrew: afarseq
Arabic: khukh
Persian: hulu
Hindi: aru
Chinese: tao
Japanese: momo
Indonesian: persik

4. Unsweetened pack. Dissolve 1 teaspoon ascorbic acid (or ascorbic acid mixture according to container directions) in 1 quart of water. Cover peaches, seal and freeze. Note: Unsweetened peaches do not have as good a color or plump a texture as those peaches frozen with sugar. The fruit freezes harder and takes longer to thaw.

5. Pectin syrup. (Peaches frozen in a pectin syrup retain their texture better than those frozen in water or juice.) Combine 1⅔ ounces of powdered pectin with 1 cup water. Heat to boiling and boil for 1 minute. Remove from heat and add 1¾ cups water. Cool. Add ½ teaspoon ascorbic acid. Low calorie or non-calorie sweetener can be added to taste if desired. Makes about 3 cups of moderately thick syrup. Add more water if thinner syrup is desired. Cover fruit with syrup, pack, seal and freeze.

CAN THEM

Peaches retain their taste and color well when canned and the payoff for the work of canning peaches is the wonderful peach dishes that can be prepared even when the fruit is out of season.

If you are selecting fruit to can, choose ripe, firm peaches. Peel and cut into halves or slices, removing the tips. To prevent the fruit from darkening during the preparation for canning, use an ascorbic acid according to manufacturer's suggestion or toss them in a mixture of two

Chilton Co. Peaches

Tablespoons lemon juice to one gallon of water.

Now you are ready to can the peaches for use later. Here are canning instructions provided by the Alabama Farmers Federation Women's Division:

1. Make a medium syrup by combining three cups of sugar and four cups of water in a saucepan. Heat until the sugar dissolves. Keep syrup hot until needed. Heat peaches thoroughly in hot syrup.

2. Pack peaches into jars, leaving ½-inch head space. Cover with boiling syrup, making sure to maintain the ½-inch head space.

3. Adjust lid, process pints for 20 minutes and quarts for 25 minutes in boiling water bath.

OR

1. Pack raw peaches into jars, leaving ½-inch head space.

2. Adjust lids and process pints for 25 minutes and quarts for 30 minutes in boiling water.

PICKLE THEM

Pickled peaches, with their tangy sweet-sour taste, can be used in salads, eaten by themselves or used to accompany almost any meal.

Here's a suggestion for pickling peaches. It, too, is provided by the *Alabama Farmers Federation Women's Division*:

3 quarts sugar
2 quarts vinegar
7 2-inch pieces stick cinnamon
2 Tablespoons cloves, whole
16 pounds (about 11 quarts) peaches, small or medium size

> King Louis XIV was a glutton in respect to every food he liked. It is said that when he was served peaches, he was so impatient to eat them that he couldn't wait for them to be peeled and bit directly into the fruit. He once granted a hefty pension to a man who furnished the royal table with peaches from Montreuil, a Paris suburb which - at the time - raised some of the finest peaches in France.

Chilton Co. Peaches

1. Combine sugar, vinegar, cinnamon and cloves. (Cloves may be put in a clean cloth, tied with a string and removed after cooking if not desired in the packed product.) Bring to a boil and let simmer, covered, for about 30 minutes.

2. Wash peaches and remove skins. To prevent pared peaches from darkening during preparation, immediately put them into cold water containing 2 Tablespoons each of salt and vinegar per gallon. Drain just before using.

3. Add peaches to the boiling syrup, enough for 2 or 3 quarts at a time and heat for about 5 minutes. Pack hot peaches into clean, hot jars. Continue heating in syrup and packing peaches until all peaches are done. Add 1 piece of stick cinnamon and 2 to 3 whole cloves (if desired) to each jar. Cover peaches with boiling syrup to ½ inch of top of jar. Adjust jar lids.

4. Process in boiling water for 20 minutes (Start to count processing time after water in canner returns to boiling.) Remove jars and complete seals, if necessary. Set jars upright, several inches apart, on a wire rack to cool.

Yield: 7 quarts.

What to can, what to freeze

Peaches from Chilton County and other parts of Alabama can be savored right away or saved for later. There are some suggestions, from the *Chilton County Extension Center*, about the best way to enjoy various Alabama varieties. *(Peach varieties are listed in order of ripening, with the earliest ones first.)*

Best Ways to Enjoy Alabama Varieties of Peaches

Variety   Type of Use

Variety	Fresh	Freeze	Pickle	Can
Springold	X		X	
June Gold	X		X	
Dixired	X		X	
Maygold	X		X	
Sentinel	X	X	X	
Coronet	X	X	X	
Redhaven	X	X		X
Harvester	X	X		X
Ranger	X	X		X
Red Globe	X	X		X
Loring	X	X		X
Blake	X	X		X
Dixiland	X	X		X
Redskin	X	X		X
Elberta	X	X		X
Jefferson	X	X		X
Rio-oso-gem	X	X		

Chilton Co. Peaches

6

Celebrating Chilton County's peaches

Each year, Chilton County celebrates the peach harvest with a Peach Festival. The festival is held on the third weekend in June.

Festival events include arts and crafts, a parade, peaches and peach recipes and the crowning of each year's "Miss Peach." There also is a peach auction during which area banks bid - sometimes as high as $1,500 - for a basket of Chilton County peaches. The proceeds from the auction go to charity. The basket of peaches traditionally is presented to Alabama's Governor as a gift from the winning bidder and Chilton County.

> The Chinese believe that eating a peach on New Year's will bring immortality.

Chilton Co. Peaches

7
Cooking with peaches

BREAKFAST DISHES

Baked Peaches

Peel and halve 6 peaches. Place in shallow pan. Fill each cavity with 1 teaspoon sugar, ½ teaspoon butter, a few drops of lemon juice and a pinch of nutmeg. Cook 20 minutes in slow oven. Serve on buttered toast.

...*Lucile Atkins*

Chilton Co. Peaches

BREADS

Peach Bread

½ cup butter or margarine, softened
1 cup sugar
3 eggs
2¾ cups all-purpose flour
1½ teaspoons baking powder
½ teaspoon baking soda
1 teaspoon salt
1½ teaspoons ground cinnamon
2 cups sliced fresh peaches
3 Tablespoons frozen orange juice concentrate, thawed and undiluted
1 teaspoon vanilla extract

    Cream butter; gradually add sugar, beating well. Add eggs, one at a time, beating well after each addition.
    Combine next 5 ingredients; add to creamed mixture alternately with the peaches, beginning and ending with flour mixture. Stir in orange juice concentrate and vanilla.
    Pour batter into greased and floured 9" x 5" x 3" loaf pan. Bake at 350 degrees for 1 hour or until wooden pick inserted in center comes out clean. Cool in pan 10 minutes; remove from pan and cool completely.

Yield: 1 loaf

...*Annette Lathan*

@@@

Peach Oat Bran Muffins

2 cups Quakers Oat Bran cereal
¼ cup firmly packed brown sugar
2 teaspoons baking powder
¼ teaspoon salt
1 cup peaches, chopped very fine

2 egg whites
¼ cup honey
2 Tablespoons vegetable oil

    Heat oven to 425 degrees. Using food processor, process cereal until it is the consistency of flour. Combine all dry ingredients. Add peaches, egg whites, honey and oil. Mix just until dry ingredients are moistened. Fill prepared muffin cups almost full. Bake 15 to 17 minutes.

Yield: 1 dozen muffins

...Brenda Elliott

CASSEROLES

Peachy Bean Casserole

1 can (16-ounce) brown sugar beans
2 Tablespoons tomato catsup
¼ cup peach preserves
2 Tablespoons chopped onion
1 Tablespoon soy sauce
4 chicken thighs or breasts

    Combine beans, catsup, preserves, onion and soy sauce in a 10" x 6" x 2" baking dish. To coat chicken pieces evenly with sauce, nestle chicken in bean mixture, skin-side down, then turn pieces skin-side up. Cover and bake in preheated 350-degree oven for 1 hour. Uncover and bake 30 minutes longer, basting chicken with sauce occasionally.

Serves: 4

...Ann Wilson

DESSERTS

Cake

Homemade Peach Cake

2 cups self-rising flour, divided
1 teaspoon salt
3 eggs, well beaten
1¾ cups sugar
1 teaspoon vanilla flavoring
1 cup vegetable oil
2 cups sliced peaches
¼ cup chopped pecans
Whipped cream

    Combine 1½ cups flour and salt; mix well and set aside. Combine eggs, sugar, vanilla and oil; beat until smooth. Add flour mixture; beat at low speed of electric mixer just until blended (batter will be thick). Dredge peaches and pecans in remaining ½ cup flour; gently fold into batter. Spoon batter into a greased and floured baking pan. Bake at 325 degrees for 1 hour. Cool completely. Cut into squares to serve. Garnish each square with a serving of whipped cream.

Note: Recipe also is delicious when served with homemade peach ice cream.

...Annette Sunday

Luscious Peach Cake

1 box yellow cake mix
1 3-ounce box of peach Jello
1 teaspoon vanilla
½ cup milk
⅔ cup cooking oil
4 eggs
2 cups coarsely cut peaches
½ cup coconut
½ cup pecans

    Sift cake mix and jello, add vanilla, milk and oil. Mix until dry ingredients are moist. Add eggs one at a time, beating well. Fold in peaches, coconut and nuts.
    Pour cake into 9" layer pans (mix will fill 3 pans). Bake at 350 degrees for 35 to 40 minutes or until cake springs back slightly when touched.

Topping:

1 cup sugar
1 cup sour cream
1 8-ounce container Cool-Whip

    Mix together and spread on layers of cake. Sprinkle with additional coconut and pecans.

...Jane Jones

Chilton Co. Peaches

Peach Marshmallow Refrigerator Cake

1 Tablespoon unflavored gelatin
¼ cup cold water
⅓ cup butter, softened
1 cup Confectioner's sugar
2 eggs, separated
½ pound marshmallows, cut into small pieces
4 cups sliced peaches
2 cups vanilla wafer crumbs

    Soften gelatin in water for 5 minutes. Cream butter, add sugar and blend in egg yolks. Cook over low heat, stirring constantly until thickened. Remove from heat, add gelatin and stir until dissolved. Cool slightly, add marshmallows, blend lightly and chill until mixture begins to thicken. Fold in sliced peaches and beaten egg whites. Arrange alternate layers of cookie crumbs and peach filling in a mold, beginning and ending with cookie crumbs. Chill until firm, unmold and serve with whipped cream if desired.

Serves: 8

...*Hazel Thomas*

@@@

Peach Pie Filling Cake

2 cups sugar
2 eggs
1½ cups Wesson Oil (buttery flavor is best)
1 can peach pie filling
3 cups flour (measured after sifting)
1½ teaspoons soda
1 teaspoon salt
1 teaspoon cinnamon
2 teaspoons vanilla

    Cream together the sugar, eggs and oil on medium

speed. Add peach pie filling and blend on low speed. Add soda, salt and spices to flour. Add flour mixture to peach pie filling mixture about a third at a time, blending well. Add vanilla. Grease and flour a large Bundt pan or tube pan. Pour mixture into pan and bake for 1 hour at 300 degrees. (May take 5 or 10 more minutes baking time, depending on oven.) To serve, add a dab of Cool-Whip, if desired.
Note: Very moist, freezes well.

...Myra Williams

Peach Upside Down Cake

½ cup butter
⅔ cup packed brown sugar
1¼ pounds peaches (about 4 peaches)
1½ cups cake flour
1¾ teaspoons baking powder
¼ teaspoon salt
½ cup granulated sugar
1 egg, beaten
¼ teaspoon almond extract
⅓ cup milk

Melt 3 Tablespoons butter in skillet (use a skillet with a handle that will not burn in the oven). Add brown sugar and beat slowly until sugar and butter are well blended. Peel peaches, cut in half. Arrange peaches, cut side down, in butter-sugar mixture in skillet. Sift flour, measure, add baking powder and salt and resift 3 times. Cream remaining butter; blend in granulated sugar; cream thoroughly. Add egg and beat until light and fluffy; stir in almond extract. Add dry ingredients and milk alternately in 3 or 4 portions. Pour batter over peaches and bake in moderate (350-degree) oven about 40 minutes or until cake springs back when touched with finger. Cool 10 minutes in skillet, then turn out on large serving plate. Serve warm or cool.

...Mrs. Odett Chappell

Sand Bucket Cake

1 8-ounce package cream cheese
1 cup Confectioner's sugar
1 large (12-ounce) container Cool-Whip
2 packages instant vanilla pudding
2 packages butter-type cookies (Lorna Doone, vanilla wafers, etc.)
3 cups diced peaches

    Mix cream cheese, Confectioner's sugar and Cool-Whip. Mix instant pudding as directed on box. Crush cookies very fine. Layer in large container (plastic sand-bucket with shovel) beginning with pudding, then peaches, then cream cheese mixture, repeat, finishing with crushed cookies. Decorate with sea shells, etc. Chill.
(Note: May use chocolate sandwich cookies for "Dirt Cake" appearance.)

...Ann Littleton

Cheesecake

Classic Chiffon Cheesecake

¼ cup butter or margarine
1 cup graham cracker crumbs
¼ cup sugar
2 (8-ounce) packages cream cheese, softened
1 can sweetened condensed milk
3 eggs, separated
¼ cup lemon juice
¼ teaspoon salt
Fruit topping (use fresh, frozen or canned peaches, drained, or peach pie filling)

Preheat oven to 375 degrees. In small saucepan, melt butter; stir in crumbs and sugar. Pat crumbs firmly on bottom of 9-inch springform pan or 9" x 9" baking pan. Bake 7 minutes. Cool while preparing filling. Reduce oven temperature to 300 degrees.

In large mixer bowl, beat cream cheese until light and fluffy. Add sweetened condensed milk and egg yolks; beat until smooth. Stir in lemon juice; set aside. In small bowl, beat egg whites with salt to soft peaks. Fold into sweetened condensed milk mixture; turn into prepared pan. Bake 50 to 55 minutes or until cake springs back when lightly touched about 1 inch from center. Cool slightly; chill. Remove rim of pan.

Top with peaches or peach pie filling.
Refrigerate any leftover cake after serving.

...Cora Sue Barnes

Chilton Co. Peaches

Peach King Cheesecake

20 ounces ricotta cheese
2 8-ounce packages cream cheese, softened
1½ cups sugar
5 eggs
5 Tablespoons flour
1 16-ounce carton sour cream
2 teaspoons vanilla flavoring
3 cups chopped peaches
1½ cups sugar
2 Tablespoons cornstarch
¼ cup water

    Combine cheeses with sugar; cream until smooth. Alternately add eggs, flour and sour cream a little at a time, beating at low speed with electric mixer until smooth. Add vanilla. Pour into greased springform pan; bake at 400 degrees for 15 minutes. Reduce heat to 350 degrees; bake for 45 minutes longer. Turn oven off; leave cheesecake in oven for 1 hour. Combine peaches and 1½ cups sugar in saucepan. Simmer over low heat for 15 minutes. Combine cornstarch and water; add to peach mixture. Cook and stir until very thick. Cool; pour over top of cooled cheesecake. Chill before serving.

Yield: 16 servings

...Janis Wilson

Many consider the cobbler pie the supreme peach dish. Cobblers are not all the same. To illustrate the point, a dozen of our favorites follow:

Cobbler

Favorite Lazy Cobbler Peach Pie

1 stick oleo
1 cup plain flour
1 teaspoon baking powder
1 cup sugar
⅔ cup sweet milk
1 quart peaches, sweetened

Melt oleo in bottom of pan. Mix flour, baking powder, sugar and milk. Pour in the bottom of the pan; add peaches and their juice. Bake in 350-degree oven for at least 45 minutes.

...Mrs. V.C. McKee

Chilton Co. Peaches

Good Ole Peach Cobbler

4 cups ripe, fresh, sliced peaches
1 cup sugar
½ cup butter
1½ cups all-purpose flour
½ teaspoon salt
½ cup shortening
¼ cup water

On top of stove, cook peaches, sugar and butter together until peaches are tender and the mixture thickens. Lightly butter a 10" x 6" x 2" baking pan and pour the peach mixture into it.
Combine flour, salt and shortening, cutting in shortening until it resembles coarse meal. Sprinkle water over the flour mixture, stirring it with a fork until the flour is moistened. Shape into a ball and roll out to a thickness of ⅛ inch. Cut into 1-inch strips and arrange over peaches in a lattice design. Use only half the strips, reserving the other half for later. Bake for 35 minutes at 350 degrees. Take out of oven and press the crust down into the peaches. Now lattice the remaining strips over the peaches and return to the oven for 40 minutes.

@@@

Peach Cobbler

1 cup plain flour
1 cup sugar
⅓ cup baking powder (I know it seems like a lot, but this is the right amount)
1 cup milk
½ stick margarine, melted
4 cups fresh sliced peaches with 1½ cups sugar added

Mix together dry ingredients. Beat in milk and margarine. Pour into a 9" x 13" pan. Distribute peaches over all. Bake at 450 degrees for 25 minutes or until brown.

...*Ann Martin*

@@@

Peach Cobbler

2 cups sliced peaches, separated
⅔ cup sugar, separated
1 stick butter or margarine, separated
2 frozen pie crusts

Layer 1 cup peaches, ⅓ cup sugar, ½ stick butter or margarine and 1 pie crust.
Repeat above step. Sprinkle a little sugar over crust and bake 35 to 40 minutes at 350 degrees.
Serve plain or with Cool-Whip

...*Bess Bentley*
First Place Winner
Pie Division
1985 Peach Cookoff

@@@

Peach Cobbler

In a buttered baking dish, pour 1 can peaches. Sprinkle with ½ cup sugar. Mix ½ box of Betty Crocker yellow cake mix with ½ cup water and dot over peaches. Cut up 1 stick butter and dot over top. Sprinkle with nutmeg and sugar on top. Cook in 350-degree oven for 30 to 45 minutes.

...*Ann Wilson*

Easy Peach Cobbler

3 cups sliced fresh peaches
1 teaspoon lemon juice
½ teaspoon cinnamon
½ teaspoon nutmeg
1 cup self-rising flour
1 egg, slightly beaten
1 cup sugar
1 stick butter or margarine, melted

Place peaches in buttered, oven-proof baking dish. Sprinkle with lemon juice, cinnamon and nutmeg. Mix flour, egg and sugar to a coarse crumb texture. Spread over peaches. Pour melted butter evenly over flour mixture. Bake uncovered at 350 degrees for 30 to 35 minutes.

Serves 6

...*Alabama Farmers Foundation Women's Division*

Easy Peach Cobbler

1 stick butter
1 cup sugar
¾ cup flour
2 teaspoons baking powder
Pinch of salt
¾ cup milk
1½ pint peaches, sliced
½ cup milk
3 cups fresh sliced peaches
1 cup fresh blueberries
½ cup sugar

Put butter in very deep baking dish and place in 350-degree preheated oven. Make a batter of the sugar, flour, baking powder, salt and milk. When batter is mixed, pour over butter. do not stir. Pour sliced peaches into batter. Do not stir. Bake at 350 degrees for 1 hour.

...*Lucile Veazey*

Lazy Day Peach Cobbler

1 stick margarine
1 cup plain flour
1 teaspoon baking powder
1 cup sugar
1 cup milk
2 to 3 cups peaches, with juice

Melt margarine in a deep baking dish. In a separate bowl, combine flour, baking powder and sugar. Add milk and mix. Pour batter over butter in baking dish. Spoon peaches over batter. Do not stir. Bake at 350 degrees for about 1 hour.

Note: This cobbler has a cake-like crust rather than a pastry-type crust.

No-Dough Peach Cobbler

½ cup butter or margarine
1 cup all-purpose flour
¾ cup sugar
2 teaspoons baking powder

Melt butter in a 2½ quart baking dish. Set aside. Combine flour, ¾ cup sugar and baking powder; add milk and stir until blended. Spoon batter over butter in baking dish; do not stir.

Combine peaches and blueberries and ½ cup sugar. Spoon over batter. Do not stir. Bake at 350 degrees for 45 to 55 minutes.

Yield: 6 servings

...Pat Conlee

Supreme Peach Cobbler

8 cups sliced, fresh peaches
2 cups sugar
2 to 4 Tablespoons all-purpose flour
½ teaspoon ground nutmeg
1 teaspoon almond extract
⅓ cup melted butter or margarine
Pastry for a double-crust, 9-inch pie

Combine peaches, sugar, flour and nutmeg; set aside until syrup forms. Bring peaches to a boil and cook over low heat 10 minutes or until tender. Remove from heat: add almond extract and butter, stirring well.

Roll out half of pastry to ⅛-inch thickness on a lightly floured board; cut into a 10" x 8" rectangle. Spoon half of peaches into a lightly buttered 10" x 8" baking dish; top with pastry. Bake at 475 degrees for 12 minutes or until pastry is golden brown. Spoon remaining peaches over baked pastry.

Roll out remaining pastry and cut into ½-inch strips; arrange in lattice design over peaches. Return to oven for 10 to 15 minutes or until lightly browned.

Yield: 8 to 10 servings

@@@

Peach Cobbler Pie

1 quart peaches, peeled and sliced
1 heaping teaspoon butter
1½ cups sugar
1 teaspoon vanilla
Batter topping (recipe below)

Cook peaches until tender in medium-large cobbler pie pan. Add butter, sugar and vanilla. Remove from heat and pour batter topping over top of peaches. Bake at 400 degrees until crust is brown and no longer sticky.

Batter Topping

1 cup flour
1 cup sugar
1 heaping teaspoon butter
Milk

Mix all ingredients with fingers until well blended. Add enough milk to make mixture the consistency of cake batter. Mix well and pour over peaches before baking.

...Myrtice Smith

Peach Roll Cobbler

1 quart fresh peaches
1½ cups sugar
1½ sticks oleo
Self-rising flour

Combine peaches, sugar and oleo and divide into three parts. Using self-rising flour, make up dough much shorter than for biscuits. Divide dough into three parts and roll each out to the size of a dinner plate. Put ⅓ of peach mixture on each piece of dough; roll up. Put 3 peach rolls in deep casserole dish. Put a little butter on top of each roll. Bake at 300 degrees until nice and brown (about 1 hour).

...Daisy Oden

Ice Cream

Fresh Peach Ice Cream

6 to 8 medium peaches, peeled and mashed
2 cups sugar
2 (13-ounce) cans evaporated milk
1 (12-ounce) can apricot nectar

Combine all ingredients in large bowl and blend well. Pour into ice cream freezer container and freeze. Ripen at least one hour before serving.

Yield: 1 gallon

Fresh Peach Ice Cream (Cooked Custard)

2 cups sugar
2 Tablespoons all-purpose flour
½ teaspoon salt
3 eggs, beaten
4 cups milk, divided
1 cup whipping cream
1½ teaspoons vanilla
6 cups mashed fresh peaches
1 cup sugar

Combine 2 cups sugar, flour and salt; add eggs and blend well. Add 2 cups of milk and cook slowly over low heat until slightly thickened; cool. Add whipping cream, remaining 2 cups of milk, vanilla and peaches which have been sweetened with 1 cup of sugar. Pour into 4-quart container of electric or hand-turned ice cream freezer and freeze.

Yield: 1 gallon

...Indiana Henry

Homemade Peach Ice Cream

2 cups mashed fresh peaches (Select soft peaches. Directions for mashing below)
¼ cup sugar
2 eggs
Pinch of salt
2 Tablespoons sugar
1 teaspoon vanilla flavoring
½ pint cream

Peel, slice and mash (either by hand or with electric blender) peaches. Add ¼ cup sugar to peaches and stir until dissolved. When sugar is thoroughly dissolved, pour into tray and freeze for 45 minutes to 1 hour. Separate eggs and beat egg yolks slightly. Add salt to egg whites and beat until stiff. Beat 2 Tablespoons sugar and vanilla flavoring into egg whites. Whip cream and fold into beaten egg mixture. Fold cream mixture and egg whites into egg yolk mixture. Fold in frozen peach mixture. Return to freezer unit to freeze.

Note: May use peach pieces in ice cream.

...Annette Sunday
First Place Winner
Ice Cream Division
1985 Peach Cookoff

Homemade Peach Ice Cream

6 eggs
¾ cup sugar
1 can sweetened condensed milk
1 Tablespoon vanilla
3 cups fresh peaches (peeled, mashed and sweetened to taste)
Homogenized milk

Beat eggs and sugar together with electric mixer; add sweetened condensed milk, vanilla and mashed peaches and fill ice cream freezer container. Finish filling container with homogenized milk within 3 or 4 inches of top. Freeze in either electric or crank-type freezer.

Yield: 6 quarts

...Blanche Dennis

Peach Ice Cream

1 pint milk
1 cup heavy cream
1½ cups sugar
1/3 teaspoon almond extract
2 cups peach pulp

Scald milk and cream, add sugar and allow the mixture to cool; put in the flavoring and half freeze. Add peach pulp and finish freezing. Remove the dasher, pack the ice cream down and cover closely. Set aside to "ripen" before serving. When the ice cream is served, garnish with a sprig of mint or a sprinkling of nutmeg.

...Alabama Farmers Foundation Women's Division

Peach Ice Cream

6 peaches
2 cups sugar
1 can Eagle Brand Milk
1 large can Pet Milk
1 can Nehi Peach soda
½ gallon whole milk
1 Tablespoon vanilla

Cut up peaches and process in blender. Add other ingredients. Pour into ice cream freeze container and freeze until hard.

...Ann Littleton

Peach Ice Cream

6 eggs
2 cups sugar
2 large cans Pet Milk
1½ teaspoons vanilla
4 cups (8 or 10) peaches
Milk

Beat eggs. Gradually add sugar. Blend well. Add Pet Milk. Beat. Add vanilla. Cut up peaches and process in blender. Add to mixture. Pour into ice cream freezer container and fill to the line with milk. Freeze until hard.

...Pat Conlee

Pie

Peach Cream Pie

Crust: ½ cup butter
        1½ cups flour
        ½ teaspoon salt

Filling: 4 cups fresh, sliced peaches
        1 cup sugar, divided
        2 Tablespoons flour
        1 egg
        ¼ teaspoon salt
        ½ teaspoon vanilla flavoring
        1 cup sour cream

Topping: 1 teaspoon cinnamon
        ⅓ cup sugar
        ⅓ cup flour
        ¼ cup butter

Crust: Cut butter into flour and salt. Press dough into 9-inch pie pan.

Filling: Slice peaches into a bowl; sprinkle with 1/4 cup sugar. Let stand while preparing rest of filling. Combine 3/4 cup sugar, flour, egg, salt and vanilla. Fold in sour cream. Stir into peaches. Pour into crust.

This pie is baked at two different temperatures. Set oven at 400 degrees and bake for 15 minutes. After you've baked the pie for 15 minutes at 400 degrees, reset oven to 350 degrees. Bake pie for another 20 minutes.

While pie is baking, prepare:

Topping: Combine sugar, cinnamon, flour and butter. Mix until crumbly.

After the 20-minute baking period at 350 degrees, remove pie from oven and sprinkle topping evenly over the top of the pie or around the edges. Bake the pie for another 10 minutes at 400 degrees.

...Indiana Henry

Peach Pie (Dutch Style)

10 to 12 ripe peaches
1 (9-inch) unbaked pie shell
1 egg
1 cup sour cream
¼ teaspoon salt
¾ cup sugar
2 Tablespoons all-purpose flour
½ teaspoon ground cinnamon
½ teaspoon ground nutmeg
Peach Pie Topping (recipe below)

Peel peaches, slice and arrange in pie shell. Beat egg slightly; mix with sour cream, salt, sugar, flour, cinnamon and nutmeg. Pour over peaches. Bake at 350 degrees for 20 minutes. Sprinkle with Peach Pie Topping and bake 15 to 20 minutes longer.

Peach Pie Topping

¼ cup brown sugar
2 Tablespoons butter or margarine
1 cup chopped nuts
3 Tablespoons all-purpose flour

While pie is baking, mix ingredients together. Sprinkle over pie as directed above.

Yield: 1 9-inch pie

...Indiana Henry

Fresh Peach Pie

1 cup sugar
2 Tablespoons flour
1 stick oleo or ¼ pound butter
3 egg yolks
2 cups chopped peaches

    Mix all ingredients together and cook in unbaked pie shell at 350 degrees until done. Top with well beaten egg whites and brown slightly.

...Mrs. E.S. Jones

Luscious Peach Pie

2 (3-ounce) packages cream cheese, softened
¾ cup sifted powdered sugar
¼ teaspoon almond extract
¾ cup whipping cream, whipped
1 (16-ounce) can sliced peaches or 2 cups fresh peaches, sliced
1 (9-inch) graham cracker pie crust

    Combine first three ingredients; beat until smooth. Fold in whipped cream and gently stir in peaches. Pour filling into graham cracker crust. Chill well.

Yield: One 9-inch pie

Old-Fashioned Peach Custard Pie

2 cups milk
¾ cup sugar
⅓ cup flour
2 eggs, separated
¼ teaspoon salt
1 teaspoon vanilla
2 cups sliced canned peaches, drained
2 Tablespoons sugar (for meringue)
Baked pastry shell (9- or 10-inch)

Put milk in double boiler and scald. Mix ¾ cup sugar, flour and egg yolks that have been slightly beaten and add slowly to milk, stirring constantly. Cook for 15 minutes, continuing to stir. Add salt and vanilla. Add peach slices which have been re-sliced to make thinner pieces. Cool a little and spoon into pastry shell. Beat egg whites until stiff, adding the 2 Tablespoons sugar gradually. Spread over pie. Bake 15 minutes in a slow oven (325 degrees) or until meringue is browned as desired. Cool before serving.

...Myra Williams

Quick Peach Pie

1 stick oleo
¾ cup self-rising flour
1 cup sugar
1½ cups sweet milk
3 cups peaches

Melt oleo in pan. Mix together flour, sugar and milk and stir well. Pour batter over melted oleo. Spoon peaches over batter. Do not stir. Bake at 350 degrees for 1 hour.

...Gladys Roberts

Chilton Co. Peaches

Sunday Peach Pie

¾ cup butter or margarine
2 cups graham cracker crumbs
1 cup sugar
1 8-ounce package cream cheese
2 envelopes Dream Whip
Peach Glaze (recipe below)
Nuts (optional)

Melt butter and mix with graham cracker crumbs. Put half in bottom of long serving dish. Cream sugar and cream cheese. Beat Dream Whip according to directions on package. Mix cream cheese mixture and Dream Whip until smooth. Put half on top of graham cracker crumbs. Put peach glaze on top of this. Add rest of cream cheese mixture, then other half of graham cracker crumbs. Top with chopped nuts, if desired. Keep in refrigerator

Peach Glaze:
1 cup sugar
3 Tablespoons cornstarch
Pinch of cinnamon
1 cup water
2 to 3 cups fresh peaches, cut in small pieces (add fruit protector of lemon juice to keep them from turning dark)

Mix sugar, cornstarch, cinnamon and water over medium heat until thickened. Add peaches.

...Debra Pitts
First Place Winner
Miscellaneous Division
1985 Peach Cookoff

Pudding

Mama's Peach Pudding

1 large can sliced peaches
1 pound vanilla wafers
4 Tablespoons flour
1¾ cups sugar
2½ cups milk
5 eggs, separated
3 Tablespoons margarine

Layer the peaches and vanilla wafers in a deep baking dish. Combine the flour and sugar. Add the milk, egg yolks and margarine. Cook in a double boiler until slightly thick, then pour over the vanilla wafers and peaches. Beat the egg whites until stiff. Add 1 Tablespoon of sugar for each egg white. Spread over the pudding mixture and place in a medium-warm over to brown.

Serves: 6

Peach Pudding

3-4 medium peaches, sliced
¾ cup sugar
2 Tablespoons shortening
½ cup milk
1 cup flour
Pinch of salt
1 teaspoon baking powder
1 teaspoon vanilla

Butter a 9-inch glass pie plate. Fill with sliced peaches. Combine remaining ingredients and pour over peaches. Bake at 350 degrees for 30 minutes

...Ann Littleton

Easy Peach Pudding

1 package instant French vanilla pudding mix
1 small package sliced peaches
1 pie shell, baked
Maraschino cherries

Mix pudding according to package directions. Pour ½ of mixture into baked pie shell and put ½ of the peaches on top to cover. Add rest of pudding and arrange rest of peaches on top. Place cherry in center.

...*Doris Morris*

Fresh Peach Pudding

6 fresh peaches or 1 pint peaches
1 (12-ounce) box vanilla wafers
1 stick oleo
1½ cups powdered sugar
1 teaspoon vanilla
2 eggs
1 cup whipping cream or 1 (2-ounce) envelope whipped topping mix prepared according to package directions

Cook peaches until soft. Crumble wafers into fine crumbs. Put half of crumbs in bottom of 13" x 9" x 2" pan. Beat oleo, sugar, vanilla and eggs together until light and creamy. Spread over crumbs. Whip cream and spread over oleo mixture. Spread peaches over cream. Sprinkle remaining crumbs over peaches. Chill overnight in refrigerator or put into freezer until firm but not frozen.

Yield: 8 servings

...*Myrtice Smith*

Miscellaneous

Perfect Peach Parfait

2 cups (16-ounce) sour cream
1 (20-ounce) can sliced peaches or 1½ to 2 pints frozen peach slices
1 can sweetened condensed milk
¼ cup lemon juice
1 cup chopped pecans
Maraschino cherries

    In large bowl, combine sour cream, peaches, sweetened condensed milk and lemon juice; mix well. Layer with nuts in parfait glasses. Garnish with cherries. Chill. After serving, refrigerate any leftovers.

Serves: 10

Note: Dessert may be frozen in an 8" x 8" pan. Garnish with nuts and cherries. Freeze 3 hours. Remove from freezer 10 minutes before serving.

...*Cora Sue Barnes*

Peach Delight

1 cup plain flour
1 stick margarine
1 cup chopped pecans

    Mix and press into large Pyrex dish. Bake until slightly brown.
Mix: 1 8-ounce package softened cream cheese
    1 medium container Cool-Whip
    2 cups powdered sugar

    Add to top of pecan mixture

Cook: 1 cup sugar
    1 cup water
    5 Tablespoons cornstarch

    Cook until thick and clear.

Add: 5 Tablespoons apricot Jello.

    Stir, then allow to cool. Slice peaches and put on top of cream cheese mixture, pour Jello mixture over.

...Brenda Elliott

Peach Fritters

½ cup plain flour
½ teaspoon salt
2 teaspoons baking powder
1 egg, beaten
⅔ cup milk
2 teaspoons powdered sugar
Fresh, firm peaches, sliced

Combine dry ingredient. Mix egg with milk; stir in dry ingredients. Batter should be thick enough to coat fruit. Sprinkle sugar over peaches. Dip peaches into batter. Fry in deep fat (370 degrees) until peaches are tender.

...Mrs. Odett Chappell

Peach Mousse

2 cups fresh, ripe peaches, peeled and sliced
⅔ cup sugar
2 cups whipping cream, whipped very stiff
3 or 4 drops almond extract

Cover peaches with sugar and let stand 1 hour. Mash and rub through sieve. Fold in whipped cream. Add extract. Pour into refrigerator trays. Freeze without stirring. Serve on lettuce leaves.

Serves: 4

...Dorothy Moon

Chilton Co. Peaches

Peach Pizza

Crust:
2 sticks margarine
2 cups flour
1 cup nuts

    Melt margarine, add flour and nuts. Press into pizza pan. Bake at 350 degrees until brown. COOL COMPLETELY.

First Layer:
1 8-ounce package cream cheese
3 cups Confectioner's sugar
1 (12-ounce) container Cool-Whip

    Blend cream cheese and sugar until smooth; fold in Cool-Whip. Spread on crust. Peak up on sides so that the top won't run off.

Top Layer:
1 cup sugar
1 cup water
3 Tablespoons cornstarch
1 box peach-flavored Jello
2 pints peaches, cut up

    Mix sugar, water and cornstarch in pan and bring to a boil; cook until clear. Cool a little. Add peach Jello and cool completely. Add peaches and spread on top.

Note: You can use either fresh sweetened peaches or frozen peaches in this recipe.

Peach Popsicles

6 large ripe peaches
¾ cup sugar
½ cup water
2 Tablespoons lemon juice
1 package unflavored gelatin
1 cup lukewarm water
Pinch of salt

Puree peaches with sugar, ½ cup water and lemon juice in blender. Dissolve gelatin in 1 cup lukewarm water and add to blender container. Add pinch of salt. Blend until smooth. Pour into 14 small paper cups. Freeze until firm. Roll in hands, then push up from the bottom to eat.

Peach Trifle

1 package yellow or white cake mix
1 Tablespoon brandy or rum flavoring
2 cups fresh peaches, sliced
1 (3-ounce) package instant vanilla pudding
1 cup chilled whipping cream
¼ cup sugar
¼ cup toasted slivered almonds

Bake cake as directed and cube. Sprinkle brandy or flavoring over cubed cake. Spoon peaches over cake. Prepare pudding. Let set 3 minutes and pour over peaches. Beat cream with sugar. Spoon over trifle. Sprinkle with almonds

...Sandy Simmons, First Place Winner
Cake Division, 1985 Peach Cookoff

Peach Trifle

1 layer of plain cake
1 package peach Jello, made according to package directions and set
1 package vanilla pudding, made according to package directions
2 cups frozen or fresh peaches, sweetened to taste and cooked slightly to retain color
Cool-Whip

In an 8" x 8" Pyrex dish, crumble up ½ cake and dribble ½ Jello over it. Add ½ of the pudding and 1 cup of peaches. Spread Cool-Whip on top. Repeat with remaining ingredients, reserving a few peaches for garnish. Refrigerate.

...*Alice Calfee*

@@@

Quickie Peach Dessert

Slice fresh peaches, add sugar and lemon juice. Sprinkle with cinnamon. Let stand 30 minutes. Serve over sliced pound cake or dessert shells. Top with ice cream, whipped cream or topping mix.

@@@

Super Peach Crisp

4 cups peeled peach slices
⅔ cup packed brown sugar
½ cup flour
½ cup old-fashioned or quick oats, uncooked

1 teaspoon cinnamon
1 teaspoon vanilla
⅓ cup margarine
½ pound Velveeta processed cheese spread, cut into small cubes
½ cup chopped pecans

Place peaches in a 2-quart oblong Pyrex dish. Combine dry ingredients, cut in margarine until mixture resembles coarse crumbs; add vanilla. Stir in processed cheese spread and nuts; sprinkle over peaches. Bake at 350 degrees for 30 minutes. Serve with whipped cream or topping, if desired.

Yield: 12 regular servings or 9 large servings

...*Myra Williams*

JAMS, JELLIES AND PRESERVES

Peach Jam (Tart Variety)

1 pound ground peaches (directions for grinding below)
1½ cup sugar

Peel peaches and remove seeds. Grind peaches in food chopper using medium blade. Weigh fruit, add sugar. Heat slowly in a thick-bottomed saucepan and stir well. Cook rapidly until somewhat thick. Pack in hot canning jars. Adjust jar lids and bands. Process in a boiling water bath canner (212 degrees) 10 minutes (for both pints and quarts).

Peach Freezer Jam, Light

3⅓ cups crushed peaches
2 Tablespoons lemon juice
3 cups sugar
1 package Sure-Jel Light

Follow directions on Sure-Jel Light package.

...*Pat Conlee*

Peach Jam with Pectin

1 quart peeled, pitted and sliced peaches
3 to 4 teaspoons liquid artificial sweetener
1 package (1¾-ounce powdered fruit pectin
1 Tablespoon fresh lemon juice
½ teaspoon ascorbic acid powder

Crush peaches in saucepan. Stir in sweetener, pectin, lemon juice and ascorbic acid powder. Bring to a boil and boil 1 minute. Remove from heat. Continue to stir 2 minutes. Pour into sterilized half-pint canning jars with tight fitting lids. Store in refrigerator. Let jam cool about 15 minutes before putting it into the refrigerator

Yield: 2 cups (1 Tablespoon = 10 calories)

Peach Butter

6 pounds fresh peaches
¼ cup fresh lemon juice

1 teaspoon cinnamon
3¼ cups sugar

    Peel peaches, cut into fourths. Puree in blender container. Combine 11 cups peach puree, lemon juice and cinnamon in large saucepan. Mix well, cook uncovered over medium heat for 2 hours, stirring frequently. Add sugar. Mix well. Cook mixture, uncovered, until desired consistency is reached, stirring frequently. Ladle into hot sterilized jars, leaving ½-inch headspace. Seal with 2-piece lids. Process in boiling water bath for 10 minutes.

Yield: 6 8-ounce jars

Note: May substitute nectarines for peaches.
                                            ...*Evelyn Littleton*

Peach Butter

2 quarts peach pulp, made from about 1½ dozen medium-sized, fully ripe peaches (directions for making pulp below)
4 cups sugar

    To prepare pulp: Wash, scald, pit, peel and chop peaches. Cook until soft, adding only enough water to prevent sticking. Press through a sieve or food mill. Measure pulp.
    To make butter: Add sugar to pulp; cook until thick (about 30 minutes). As mixture thickens, stir frequently to prevent sticking. Sterilize canning jars. Pour hot butter into jars, leaving 1/4-inch headspace. Wipe jar rims and adjust lids. Process in boiling water bath for 5 minutes.

Yield: About 8 half-pint jars

Spiced Peach Butter

Follow above recipe for Peach Butter. Add ½ to 1 teaspoon each ground ginger and ground nutmeg with sugar to peach pulp. Process as described above.

Peach Marmalade

3 pounds peaches
3 oranges
6¾ cups sugar
1½ pints water

To prepare fruit: Wash, peel and slice peaches into very thin strips or pieces. Peel oranges and thinly slice peel. Separate seeds and membrane from orange pulp. Cut pulp into pieces.

To make marmalade: Sterilize canning jars. Boil sugar and water until sugar is dissolved. Add the fruit. Cook rapidly, stirring frequently until jellying point is reached. The fruit should appear in small pieces throughout the mixture and the mixture should be smooth in consistency. Pour hot marmalade into hot jars, leaving 1/4-inch headspace. Wipe jar rims and adjust lids. Process 5 minutes in a boiling water bath.

Yield: About 7 half-pint jars

Peach Preserves

1 pound peaches
1½ cups sugar
⅓ to ½ cup water

Wash peaches, remove stems, peel and remove pits. Cut into uniform-size pieces. Add the sugar and water and cook at once. Heat slowly to boiling, stirring constantly, then boil rapidly. Cook until syrup is somewhat thick. Pack hot in hot canning jars or let peaches stand in syrup overnight to plump up. Pack cold the next morning in warm jars, leaving ½-inch headspace. Adjust lids and screwbands. Process in a boiling water bath. (Hot pack, 10 minutes; cold pack, 25 minutes)

Yield: 2 pints

...Chilton County Extension Service

Peach Preserves

4 quarts peaches
4 cups sugar
2 cups hot water

Blanch peaches, remove skins and cut into halves. Add sugar and water to peaches, cover and let stand 2 hours. Heat slowly until sugar is dissolved and syrup is boiling. Boil rapidly until peaches are tender and clear. Let stand overnight to "plump." Bring to boiling point and pack into clean, hot jars and seal immediately. NOTE: If syrup is thin, drain it off and cook down until it has the consistency of honey. Add peaches, bring to boiling point and pack. Process 15 minutes.

...Alabama Farmers Foundation Women's Division

Chilton Co. Peaches

MAIN DISHES

Center Cut Ham with Peaches

1 16-ounce can peaches
3 to 3½ pound center cut ham slice, 2 inches thick
¼ cup brown sugar
2 Tablespoons cornstarch
2 Tablespoons vinegar
¼ teaspoon nutmeg
2 Tablespoons butter or margarine

    Drain peaches, reserving juice. Trim fat from ham slice; score edges. Place in 2-quart, glass baking dish. Combine reserved peach juice, brown sugar, cornstarch, vinegar and nutmeg in small mixing bowl. Pour over ham. Top with peaches. Dot with butter. Microwave on "Roast" from 20 to 25 minutes or until ham is hot and sauce is thickened, turning once. Let stand for 5 minutes before serving.

Yield: 6 to 8 servings

...Mae Britt

Peachy Chicken

1 jar (10 ounces) peach preserves
⅔ jar Russian dressing
1 package dry onion mix
12 pieces chicken, cut up

    Mix together preserves, dressing and onion mix. Spread over chicken in broiler pan. Bake at 350 degrees for 1½ hours.

Easy and fantastically good.

Serves: 6

...Brenda Elliott

@@@

Tenderloin with Peach Glaze

3 cups fresh peaches, mashed
2 cups water, divided
½ cup granulated sugar
1/4 teaspoon dry mustard
1 Tablespoon soy sauce
½ teaspoon salt
2 pork tenderloins, about 1 pound each
Salt and pepper

Preheat oven to 325 degrees. Place mashed peaches, 1 cup of the water and sugar in a saucepan. Bring to a boil, then reduce heat and simmer for 45 minutes. Mix mustard, soy sauce and salt into the glaze. Cook an additional five minutes. Salt and pepper the tenderloins then place in a greased baking pan with the remaining 1 cup of water. Bake 30 minutes per pound (total weight of both tenderloins) in the preheated oven. During last 30 minutes of baking time, pour glaze over the roast. Baste roast with the glaze several times during the last 15 minutes of cooking time.

...Ella Nelle Guy

Chilton Co. Peaches

PICKLES

Peach Pickles

2 gallons peaches (White clingstone peaches are preferred. Select firm fruit.)
1 quart sugar
1 quart clear, distilled vinegar
2 cinnamon sticks
1 Tablespoon whole cloves
1 Tablespoon allspice

    Wash peaches well. Remove the thin skin carefully and drop peaches at once into syrup made by cooking sugar, vinegar and spices until slightly thick. Let stand overnight. Drain off liquid and boil it again until slightly thick, then add fruit. Do not stir, but keep peaches under syrup until tender. Pack hot into hot pint canning jars. Cover with syrup. Adjust lids. Process in a boiling water bath canner for 10 minutes.

...Ann Wilson

@@@

Peach Pickles

3 pounds sugar
1 pint vinegar
1 cup water
2 ounces stick cinnamon, broken
1 ounce whole cloves
6 pounds peaches

    Combine sugar, vinegar, water and spices and boil together until clear, about 15 minutes. Add peaches, only enough at one time to fill 1 jar, and cook until tender. Pack in hot jars, cover with hot syrup and seal.

...Dixie Calfee

Pickled Peaches

6 cups sugar
1 quart distilled white vinegar
3 sticks cinnamon
1 Tablespoon whole cloves
1 Tablespoon whole allspice
1 teaspoon salt
½ teaspoon ground ginger
½ teaspoon ground mace
8 pounds (about 24) firm, fresh peaches, peeled

Combine sugar and vinegar in large saucepan or Dutch oven. Tie cinnamon, cloves and allspice in cheesecloth; add to saucepan. Stir in salt, ginger and mace. Heat to boiling, stirring constantly until sugar is dissolved. Add a portion of the peaches to boiling syrup, boil 4 or 5 minutes. Pack hot peaches into sterilized jars. Repeat with remaining peaches. Remove spice bag from syrup. Pour syrup over peaches, leaving ½-inch headspace. Adjust caps. Process 20 minutes in boiling water bath.

Yield: About 6 pints

Note: Peaches can be left whole or cut in half and pitted. If peaches are not processed, they can be refrigerated up to 3 weeks.

...Darlene Littleton

Quick Method Pickle Peaches

6½ cups sugar
1 quart cider vinegar
5 quarts peeled peaches (Use a firm, ripe, pickling peach)
1¼ teaspoons mixed pickling spices
10 cloves
5 small pieces of stick cinnamon

    Make a syrup of sugar and vinegar. Heat and add peaches. Let fruit and syrup simmer for 10 minutes. To each quart canning jar, add ¼ teaspoon mixed pickling spices, 2 cloves and 1 small piece of stick cinnamon. Add hot fruit to the jar; cover with hot syrup. Adjust jar lids and bands. Process in a boiling water bath canner (212 degrees) for 25 minutes. Let peaches "season" at least one week. To develop their best flavor, wait 6 weeks.

Sweet Pickle Peaches

6 pounds peaches (Select firm, clingstone peaches)
6 cups sugar
2 cups water
1 pint cider vinegar
4 sticks cinnamon
2 Tablespoons whole cloves
1 Tablespoon ginger

    Peel peaches and drop at once into a syrup made by boiling together 3 cups sugar and 2 cups water for 15 minutes. Cool and allow to stand for 2 or 3 hours. Drain off syrup, put vinegar and spices and remaining sugar into it and cook for 15 minutes. Add peaches and cook together for 30 minutes. Let stand overnight. Next morning, drain off syrup, cook for 20 minutes, add peaches and continue boiling for 15 minutes. Cool again and let stand at least 2 hours or overnight. Then cook again until peaches are clear and tender. Pack peaches into pint canning jars;

cover with strained syrup. Adjust jar lids and bands. Process in boiling water bath canner (212 degrees) for 10 minutes.

SOUPS, SAUCES AND SALADS

Soup

Author's note: While most of the peach recipes in this book came from the very helpful and generous people in Chilton County, who probably know better than anyone just how to cook with peaches, I hope they won't mind indulging me and letting me share my favorite peach recipe:

Cold Peach Soup

4 large ripe peaches
Boiling water to cover peaches
1 cup dry white wine (This is optional, but if you don't use it, put in another cup of water to make up for the difference.)
2 cups water
3 Tablespoons sugar (Vary this to taste)
¼ teaspoon cinnamon
3 cloves
1 pint heavy cream

Drop peaches in boiling water for 1 minute. Peel and cut up. Puree in food processor or blender. Put puree in enamel saucepan. Add wine, water, sugar and spices. Bring to a boil and simmer 10 minutes, stirring. Remove cloves and chill at least 4 hours. Add heavy cream just before serving.

Chilton Co. Peaches

Salad

Peach Delight

2 envelopes unflavored gelatin
½ cup water
¼ cup sugar
¼ cup orange-flavored liqueur
1 quart vanilla ice cream, softened
2 cups sliced fresh peaches
1 cup whipping cream

    Combine gelatin and cold water in top of a double boiler; let stand 5 minutes. Place over boiling water and cook, stirring constantly until dissolved. Remove from heat; stir in sugar and liqueur. Add ice cream and peaches, mixing well. Beat whipping cream until soft peaks form. Fold into gelatin mixture. Pour into an oiled 6 or 7 cup mold. Freeze until firm. Unmold on a platter and garnish with additional peaches. May be served with whipped topping.

Yield: 10 servings

...Gladys Stone
First Place
Salad Division
1985 Peach Cookoff

@@@

Peach Congealed Salad

1 (6-ounce) package peach Jello
2 cups hot water
1 pint crushed frozen or fresh peaches (if frozen, let thaw partially)
1 (3-ounce) package cream cheese, grated

Dissolve Jello in hot water. Add peaches and cream cheese. Pour in mold and chill until firm.

...*Mrs. Willie Keen*

Exotic Peach Salad

1 cup miniature marshmallows
1 small can crushed pineapple
1 can coconut
1 carton sour cream
Peach halves

Mix marshmallows, pineapple, coconut and sour cream together. Spoon into peach halves and serve.

...*Dot Moore*

Frozen Peach Salad

1 package cream cheese
⅓ cup mayonnaise
1 teaspoon lemon juice
½ teaspoon salt
½ cup whipping cream
2 cups peaches, sliced thin
½ cup chopped pecans

Combine the cream cheese, mayonnaise, lemon juice and salt. Mix well.

Whip the whipping cream. Fold it, the peaches and the pecans into the cream cheese mixture. Freeze 4 hours in a 2-quart flat casserole. Slice and serve on crisp lettuce leaves.

Serves: 8

...*Alabama Farmers Foundation Women's Division*

Peach Jello Salad

1 large package peach Jello
1 cup boiling water
1 large can crushed pineapple
1 cup peach pie filling
¾ cup Cool-Whip

 Mix Jello and boiling water; let cool. Add pineapple and pie filling. Mix well. Add Cool-Whip.

Topping:
1 (8-ounce) package cream cheese
½ cup sugar
1 (8-ounce) carton sour cream
½ cup nuts
 Mix cream cheese and sugar, then add sour cream and nuts. Spread over Jello mixture.

...Ann Wilson

@@@

Stuffed Peach Salad

1 (3-ounce) package cream cheese
2 dozen salted almonds, finely chopped
⅛ teaspoon salt
¼ teaspoon sugar
Dash of paprika
Grated rind of ½ orange
12 peach halves
6 lettuce leaves

 Mix cream cheese, almonds, salt, sugar, paprika and grated orange rind. Shape into small balls (if cheese mixture is too dry, add cream or orange juice to soften). Arrange 2 peach halves on each lettuce leaf. Place cream cheese balls in hollow of each peach half. Serve with or without dressing.

...Mrs. Odett Chappell

Hot Weather Salad

3 cups ripe peaches, cut in chunks
1 cup blueberries
2 cups watermelon, seeded and cut in chunks
3 bananas, sliced in rounds
1 cup cantaloupe, cut in chunks
1 cup strawberries, sliced

    Combine all fruits in a large bowl. Mix gently to prevent mashing.

Peach Delight Salad

½ cup sugar
2 cups fresh, sliced peaches
1 (3-ounce) package peach Jello
1 cup boiling water
½ pound marshmallows
½ cup milk
1 cup whipping cream

    Sprinkle sugar over peaches; let stand 30 minutes. Dissolve Jello in boiling water. Drain juice from peaches; add enough water to juice to make 1 cup; add to gelatin and chill until partially set. Combine marshmallows and milk in top of double boiler, stir until marshmallows melt. Cool thoroughly and fold in whipped cream. Add peaches to gelatin and fold in marshmallow mixture. Pour into 1-quart mold and chill.

Yield: 6 servings

...Myrtice Smith

Sauce

Fresh Peach Sauce

2 cups crushed fresh peaches
½ cup sugar
½ cup orange juice
2 teaspoons lemon juice
1 teaspoon vanilla flavoring

    Combine peaches, sugar and orange juice in 1-1/2 quart saucepan. Stir over high heat under mixture reaches boiling point. Simmer, uncovered, about 15 minutes or until thick. Remove from heat and stir in lemon juice and vanilla. Cool.

...Darlene Littleton

Peach Sauce for Roasted Chicken

1 can (29 ounces) cling peaches, drained
2 teaspoons butter, melted
½ cup white wine
¼ teaspoon cinnamon
¼ teaspoon allspice
1 broiler-fryer chicken, whole
⅛ teaspoon salt
⅛ teaspoon pepper

    In blender, puree drained peaches. Slowly add melted butter, wine, cinnamon and allspice. Pour pureed ingredients into small saucepan and heat over medium heat until sauce bubbles. Brush warm sauce over entire broiler-fryer, including inside cavity. Sprinkle inside cavity with salt and pepper. Roast in 350-degree oven for about 1 hour, basting generously every 15 minutes with sauce. Serve piping hot with remaining peach sauce.

Serves: 4 to 6

...Lillian B. Leach

GENERAL INDEX

ANTI-DARKENING SOLUTIONS 20

CHILTON COUNTY ....
   Peach Festival 27
   Peach history 1

PACKS
   Canning 22
   Sugar, freezing 19
   Syrup, light, freezing 20
   Syrup, freezing 20
   Syrup, pectin, freezing 20
   Unsweetened, freezing 20

PEACHES
   Canning 22
   Choosing 13
   Freezing 19
   History 1
   Measurements 17
   Nutritional value 17
   Peeling 15
   Picking 14
   Pickling 21
   Pitting, sliced 15
   Pitting, whole 15
   Ripening 14
   Uses of varieties 7
   Varieties 7
      Babcock
      Belle of Georgia
      Biscoe
      Blake
      Burbank July Elberta
      Camden
      Com-Pact RedHaven
      Coronet
      Cresthaven
      Cumberland
      Culinan
      Dawne
      Delp Early Hale
      DesertGold
      Dixired
      Early RedFe
      Early RedHaven
      Elberta
      Garnet Beauty
      GloHaven
      Golden Jubilee

Chilton Co. Peaches

Golden Monarch
Harbinger
Harken
Harvester
Havis
J.H. Hale
Jerseyland
Junegold
Loring
Madison
Marsun
Maybelle
Monroe
Norman
Ranger
Raritan Rose
Redglobe
RedHaven
Redskin
Reliance
Rio-Oso-Gem
Sentinel
Springcrest
Stark Autumn Gold
Stark Earliglo
SunHaven
Sunhigh
Sunshine
Topaz
Triogem
Velvet
Washington
Waverly
Winblo
Yakima Hale

RECIPE INDEX

BREADS 30
Muffins, Peach Oat Bran
Peach Bread

BREAKFAST DISHES 29
Baked Peaches

CASSEROLES 31
Peachy Bean Casserole

DESSERTS 32
Cake
Homemade Peach Cake
Luscious Peach Cake
Peach Marshmallow Refrigerator Cake
Peach Pie Filling Cake
Peach Upside Down Cake
Sand Bucket Cake

Cheesecake 36
Classic Chiffon Cheesecake
Peach King Cheesecake

Cobbler 39
Favorite Lazy Cobbler Peach Pie
Good Ole Peach Cobbler
Peach Cobbler
Peach Cobbler, Easy
Peach Cobbler, Lazy Day
Peach Cobbler, No-Dough
Peach Cobbler, Supreme
Peach Cobbler Pie
Peach Roll Cobbler

Ice Cream 46
Fresh Peach Ice Cream
Fresh Peach Ice Cream (Cooked Custard)
Homemade Peach Ice Cream
Peach Ice Cream

Pie 50
Peach Pie, Cream
 Peach Pie, Dutch Style
Peach Pie, Fresh
Peach Pie, Luscious
Peach Pie, Old-Fashioned Custard

Chilton Co. Peaches

Peach Pie, Quick
Sunday Peach Pie

Pudding 55
Mama's Peach Pudding
Peach Pudding
Peach Pudding, Easy
Peach Pudding, Fresh

Miscellaneous 57
Parfait, Perfect Peach
Peach Delight
Peach Fritters
Peach Mousse
Peach Pizza
Peach Popsicles
Peach Trifle
Quickie Peach Dessert
Super Peach Crisp

JAMS, JELLIES AND PRESERVES 63
Jam, Peach, Tart Variety
Jam, Peach Freezer, Light
Jam, Peach with Pectin
Peach Butter
Peach Butter, Spiced
Peach Marmalade
Peach Preserves

MAIN DISHES 68
Chicken, Peachy
Ham, Center Cut with Peaches
Tenderloin with Peach Glaze

MISCELLANEOUS
Peach Chutney
Peach "Go-With"
Peach Leather
Peach Peelings, Frozen

PICKLES 70
Peach Pickles
Pickle Peaches, Quick Method
 Pickled Peaches
Sweet Pickle Peaches

SOUPS, SAUCES AND SALADS 73
Soup 73
Cold Peach Soup

· 82 ·

Salad 74
Peach Delight
Peach Salad, Congealed
Peach Salad, Exotic
Peach Salad, Frozen
Peach Salad, Jello
Peach Salad, Stuffed
Salad, Hot Weather
Salad, Peach Delight

Sauce 78
Sauce, Fresh Peach
Sauce, Peach for Roasted Chicken

Chilton Co. Peaches